Workbook to Accompany

FIREFIGHTING STRATEGIES AND TACTICS

James Angle
Michael Gala
David Harlow
William Lombardo
Craig Maciuba

DELMAR
™
THOMSON LEARNING

Australia Canada Mexico Singapore Spain United Kingdom United States

DELMAR

™

THOMSON LEARNING

Workbook to Accompany Firefighting Strategies and Tactics
by
James Angle, Michael Gala, David Harlow, William Lombardo, Craig Maciuba

Business Unit Director:
Alar Elken

Executive Editor:
Sandy Clark

Acquisitions Editor:
Mark W. Huth

Development:
Dawn Daugherty

Executive Production Manager:
Mary Ellen Black

Production Editor:
Ruth Fisher

Production Coordinator:
Ruth Fisher

Art/Design Coordinator:
Rachel Baker

Executive Marketing Manager:
Maura Theriault

Channel Manager:
Mona Caron

Marketing Manager:
Kasey Young

For permission to use material from this text or product, contact us by
Tel   (800) 730-2214
Fax   (800) 730-2215
www.thomsonrights.com

Library of Congress Cataloging-in-Publication Data

ISBN: 0-7668-1345-2
CIP: 00-043156

NOTICE TO THE READER

Contents

Preface

ABOUT THIS WORKBOOK

This workbook has been designed to reinforce and complement the textbook *Firefighting Strategies and Tactics*. As stated in the Preface of the textbook, fireground strategy and tactics are the essence of fire department operations. It is therefore necessary for all members of the fire service to gain a knowledge of this subject.

This workbook is comprised of chapters that match those in the textbook itself. As such, each chapter requires the reader to write responses to key terms and to put into their own words what is presented in the chapters in the text. It is hoped that by doing this the reader will gain a good foundation and a solid understanding of *Firefighting Strategies and Tactics*.

Your instructor may find this workbook useful as a tool for assigning homework, individually or as a group, or as a classroom discussion tool. Regardless of use, it is recommended that students complete each workbook chapter at the time they complete the associated textbook chapter. The student should review the chapter objectives, study the chapter, complete the review questions and activities in the textbook, and then complete the workbook pages.

Chapter

1

History of Fire Service Tactics and Strategies

LEARNING OBJECTIVES

Upon completion of this chapter, you should be able to:

- Explain significant historical changes in fire service strategies and tactics.
- List the firefighter's role in contemporary strategy and tactics.
- Explain why it is important for firefighters to have an understanding of strategy and tactics.

Define the following Key Terms:

Personnel protective equipment

_____

Nomex

_____

Kevlar

_____

Personal alert devices (PASS)

_____

National Emergency Training Center

_____

National Fire Academy

_____

United States Fire Administration

_____

Federal Emergency Management Agency

_____

"America Burning"

_____

FIRESCOPE

_____

Siamese

_____

National Institute of Occupational Safety and Health

_____

Self-contained breathing apparatus

_____

Chord and web

_____

Gusset plate

_____

Balloon frame construction

_____

Describe how the following have impacted fire service strategies and tactics:

Equipment advances

_____

_____

Personnel protective clothing

_____

_____

Advanced and increased educational opportunities

_____

_____

Codes and standards

_____

_____

Firefighter and civilian deaths

_____

_____

Building materials and building contents

_____

_____

Architectural designs and engineering

_____

_____

Wildland, urban interface

_____

_____

New equipment and materials in vehicles

_____

_____

Hazardous chemicals

_____

_____

Describe the firefighter's role in strategies and tactics:

_____

_____

_____

_____

_____

Chapter

2

Fire Dynamics

LEARNING OBJECTIVES

Upon completion of this chapter, you should be able to:

- Compare the components of the fire triangle and the components of the fire tetrahedron.
- Define the four classes of fire.
- Describe the stages of fire growth.
- Describe the four methods of heat transfer.
- Define the terms *flashover* and *back draft*.
- Describe the concept of smoke behavior.
- Describe the relationship between fire dynamics and the application of strategies and tactics.
- Understand why tactics must be undertaken in a logical sequence.

Define the following Key Terms:

Exothermic reaction

_____

Fire triangle

_____

Fire tetrahedron

_____

Convection

_____

Radiation

_____

Conduction

_____

Direct flame impingement (contact)

_____

Rollover

_____

Flashover

_____

Thermal radiation feedback

_____

Taxpayer

_____

Strip shopping center

_____

Row house

_____

Garden apartment

_____

Facade

_____

Draw the fire triangle and explain its components.

Draw the fire tetrahedron and explain its components.

Describe the classes of fire:

Class A

_____

Class B

_____

Class C

_____

Class D

_____

Class K

_____

Describe the phases of fire growth:

_____

_____

_____

_____

_____

_____

Using Figure 2-1, Fill-in the blanks with the four different methods of heat transfer.

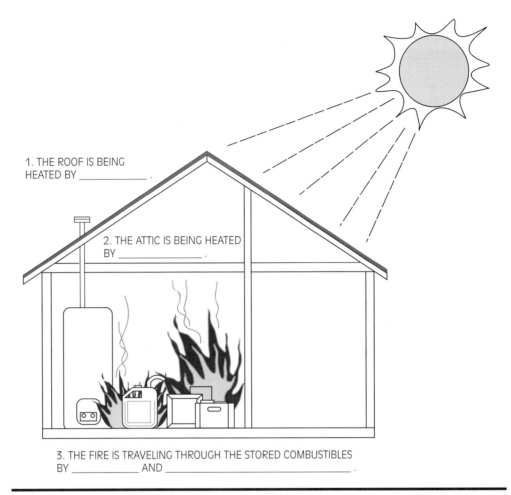

1. THE ROOF IS BEING HEATED BY _____ .

2. THE ATTIC IS BEING HEATED BY _____ .

3. THE FIRE IS TRAVELING THROUGH THE STORED COMBUSTIBLES BY _____ AND _____ .

Figure 2-1

List the Indications of flashover:

Real World Scenario: You are the Officer on a truck company. You arrive at a single-family, two-story house with flames showing out of two windows on the first floor. While the engine company stretches a handline to the fire, you search the floor above the fire. As you and your crew are searching, you notice that the heat is intense and it forces you to your knees. There is thick black smoke banking down to the 3-foot level and you notice fire rolling over your head. You are about 5 feet into the room. These are warning signs of what imminent danger? What are other signs of this danger? How do you survive this hazard?

_____

_____

_____

List flashover survival tips:

_____

_____

_____

_____

_____

_____

_____

_____

List the Indications of back draft:

Real World Scenario: You are the Officer on the unit that arrives first at a single-story house fire. On your size-up, you notice smoke churning in and out of the eaves. The windows are blackened and as you approach the house you notice that the windows are hot to the touch. The fire appears to be in the smoldering phase. What dangers are present in this type of scenario? What can be done to ensure firefighter safety? What other signs might you look for to indicate this phenomenon?

_____

_____

_____

List back draft survival tips:

_____

_____

_____

_____

_____

_____

_____

Using Figure 2-2, show possible avenues of smoke spread.

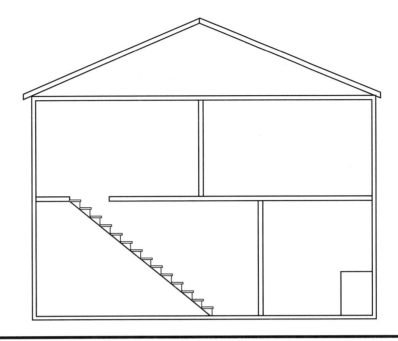

Figure 2-2

List the gases found in smoke:

_____

_____

_____

_____

_____

_____

Chapter 3

Extinguishing Agents

LEARNING OBJECTIVES

Upon completion of this chapter you should be able to:

- Discuss the various properties and use of water as an extinguishing agent.
- Describe the various formulas to determine fireground flow requirements.
- Demonstrate various types of nozzles and explain the advantages and disadvantages of each.
- Perform basic hydraulic calculations.
- Describe the use of foam as an extinguishing agent.
- Discuss the use of dry chemicals and dry powders.

Define the following Key Terms:

Normal operating pressure

_____

Surface area

_____

Subsurface injection

_____

Hydrocarbons

_____

Polar solvents

_____

Compressed air foam system

_____

Describe the properties of water and the reasons that it is a good extinguishing agent:

_____

_____

_____

_____

_____

_____

Compare the advantages, disadvantages, and usage of the following nozzles: fog, solid stream, and broken stream.

Real World Scenario: Your Company is operating a deck gun on a large tire fire. The nozzle on the deck gun is a 1000 gpm fog nozzle and you are receiving ample pressure for an adequate stream. Suddenly you experience a wind change and your stream is no longer reaching the desired target. As the Officer you direct your personnel to change to what type of nozzle? Why do you make this choice?

_____

_____

_____

Complete the following problems using the formulas presented in the text.

- Using the NFA fire flow formula, what would be the flow required for 100% involvement of a 100′ × 100′, single-story building? _____ gpm
- Using the NFA fire flow formula, what would be the flow required for 100% involvement of a 125′ × 75′, single-story building? _____ gpm
- Using the NFA fire flow formula, what would be the flow required for 50% involvement of a 200′ × 100′ building? _____ gpm
- Using the Iowa State fire flow formula, what would be the flow required for 100% involvement of a 1000 cubic foot area? _____ gpm
- Using the Iowa State fire flow formula, what would be the flow required for 100% involvement of a 10,000 cubic foot area? _____ gpm

How do the flow requirements of these two formulas compare?

Real World Scenario: You are the initial incident commander on a warehouse fire. The warehouse is fully involved upon arrival and you order a defensive operation. You quickly figure that the building is approximately 80 ft long and 40 ft wide. It is a one-story, flat-roof warehouse, so you figure it is approximately 12 ft in height. There is no one to help you presently at the command post. You need to set up accountability, safety, direct apparatus and personnel and ensure that adequate resources are en route to the scene. Quick! What is the fire flow requirement?

_____

What resources from your department or mutual aid departments will be required to meet the fire flow?

_____

- What is the flow from a 1″ tip flowing at 50 psi nozzle pressure? _____ gpm
- What is the flow from a 15/16″ tip flowing at 50 psi nozzle pressure? _____ gpm
- What is the flow from a 1¼″ tip flowing at 50 psi nozzle pressure? _____ gpm
- What is the flow from a 1¼″ tip flowing at 80 psi nozzle pressure? _____ gpm
- What is the flow from a 2″ tip flowing at 80 psi nozzle pressure? _____ gpm
- What is the friction loss in a 300′ foot lay of 2½″ hose flowing 200 gpm? _____
- What is the friction loss in a 1000′ foot lay of 5″ hose flowing 1000 gpm? _____
- What is the friction loss in a 500′ foot lay of 3″ hose flowing 500 gpm using the Q^2 Formula? _____
- What is the friction loss in a 150′ foot lay of 1½″ hose flowing 125 gpm? _____
- What is the friction loss in a 200′ foot lay of 1¾″ hose flowing 175 gpm? _____

Calculate the required pump discharge pressure for the following:

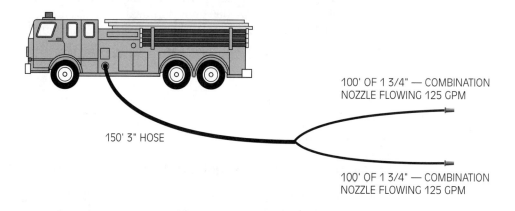

Figure 3-1   *Figure the required pump discharge pressure.*

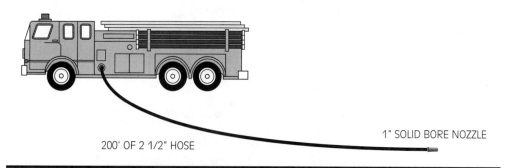

Figure 3-2   *Figure the required pump discharge pressure.*

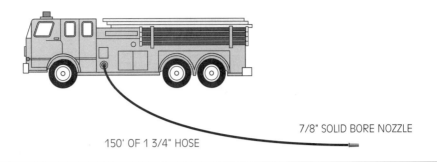

Figure 3-3 *Figure the required pump discharge pressure.*

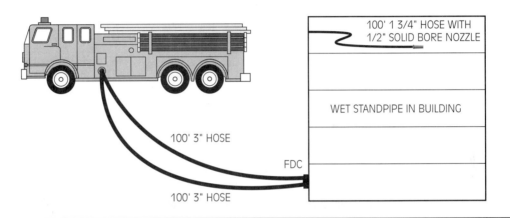

Figure 3-4 *Figure the required pump discharge pressure.*

Describe the components of foam and explain how it extinguishes a fire.

Real World Scenario: You are the Officer on an engine company who arrives first at a gasoline tanker fire. Your engine is equipped with 20 gallons of AAAF and a 95-gpm in-line eductor. You have an adjustable gallonage fog nozzle. You order your personnel to lay a fog line and place a blanket of foam on the fire. The eductor has been placed 100′ from the engine and 200′ from the nozzle. The eductor is placed in the foam concentrate and water is flowed through the nozzle. The foam blanket is very poor. What might be causing the problem?

_____

_____

_____

_____

_____

List the six types of dry chemicals:

_____

_____

_____

_____

_____

_____

Explain how dry chemicals extinguish fires:

_____

_____

_____

_____

_____

_____

Describe the following extinguishing agents and explain how they extinguish fires:

Dry powders

_____

_____

_____

Carbon dioxide

_____

_____

_____

Halon

_____

_____

_____

Chapter

4

Incident Management Systems

LEARNING OBJECTIVES

Upon completion of this chapter, you should be able to:

- Describe the need for an incident management system.
- Identify the minimum requirements of an incident management system as set forth by the National Fire Protection Association (NFPA).
- Outline the major incident management components of the Fireground Command model, the NIIMS model, and the National Fire Service Incident Management System model.
- Discuss the differences of terminology between FGC, NIIMS, and the National Fire Service Incident Management System models.
- Explain the need for and importance of standard operational policies.

Define the following Key Terms:

Span of control

_____

Sector

_____

Unified command

_____

Command

_____

Operations

_____

Planning

_____

Logistics

_____

Finance

_____

Branches

_____

Divisions

_____

Groups

_____

Describe the requirements of NFPA 1561:

_____

_____

_____

_____

_____

_____

_____

Describe the basic components of the Fire Ground Command System:

_____

_____

_____

_____

_____

Describe the relationship between strategic, tactical, and task level functions:

_____

_____

_____

_____

_____

_____

_____

_____

_____

_____

_____

Draw an IMS structure for a fire in a single-family dwelling with the response of 3 engines, 1 ladder, 1 squad, and 1 battalion chief.

Draw an IMS structure showing all command staff and general staff positions.

Compare the NIIMS, Fire Ground Command, and National Fire Service incident management systems.

_____

_____

_____

_____

_____

_____

_____

_____

_____

Describe a situation in which a unified command would be used and how it might be implemented.

_____

_____

_____

_____

_____

List the items of the IMS that should be covered in the SOGs.

_____

_____

_____

_____

_____

_____

Draw an IMS structure for the following incident. Label all divisions, sectors, or groups that you feel are required for the incident.

Real World Scenario: There is a fire on the fifth floor of a multi-story apartment building that is shown in Figure 4-1. The fire occurs at 04:00 hours, and is in one of the apartments. There is smoke throughout the fire floor and the floor above.

Figure 4-1 *A multi-story apartment building.*

Chapter 5

Coordination and Control

LEARNING OBJECTIVES

Upon completion of this chapter, you should be able to:

- Describe why effective fireground communication is necessary.
- Discuss the process necessary to ensure effective fireground communication.
- Define size-up.
- List the components of an effective size-up.
- Define operational mode.
- List and describe the three incident priorities.
- Define strategic goals.
- Define tactical objectives.
- Define tactical methods.
- Describe action planning.
- Compare the relationship of incident priorities, strategic goals, tactical objectives, and tactical methods.
- Discuss the role of preincident planning to the overall strategic and tactical plan.
- Discuss the concept of recognition-primed decision making.

Define the following Key Terms:

Communication systems

_____

Size-up

_____

REVAS

_____

RECEOVS

_____

Describe the need for effective fireground communications. What are some of the roadblocks to good communication?

_____

_____

_____

_____

_____

_____

_____

_____

Real World Scenario: You have been assigned to be the rapid intervention crew on a large bowling alley fire. Command advises you that two firefighters are missing and orders you to begin a search at their last known location. As you begin searching, you notice that fireground radio traffic is extremely busy due to the lost fire fighters and the ongoing fire fight. What options do you have to ensure that you have effective communications if they are needed?

_____

_____

_____

What actions can you take in the post incident analysis to ensure effective communications in the future?

_____

_____

_____

Describe the six steps of the communications model presented in the text.

Step 1 _____

_____

Step 2 _____

_____

Step 3 _____

_____

Step 4 _____

_____

Step 5 _____

_____

Step 6 _____

_____

Draw the size-up triangle below.

Using the table in the text as a guide, complete the following table of potential size-up factors. Use not only those in the text, but develop some of your own.

Environment	Resources	Conditions/Situation

Describe the incident priorities and give one example of how each might be accomplished.

Life safety _____

_____

_____

_____

Incident stabilization _____

_____

_____

_____

Property conservation _____

_____

_____

_____

Real World Scenario: You are the company office on a three-person engine company with a 750-gallon water tank. You arrive at the scene of a single-story, ordinary construction duplex fire with heavy smoke showing out of two bedroom windows. A hydrant is approximately 500′ from the duplex. An elderly woman is sitting in the driveway screaming that her grandchildren are in the house. You notice that the elderly woman's arm has a 3″ laceration that is bleeding profusely. She received the laceration while trying to rescue her grandchildren through a bedroom window. Several bystanders are gathering; however, your next engine is not due to arrive until at least three minutes have passed. Explain the actions you take to meet the number one fireground priority—life safety. Justify your response.

_____

_____

_____

Describe the two operational modes:

Offensive attack mode

_____

_____

_____

Defensive attack mode

_____

_____

_____

List and define the nine strategic goals presented in the text:

1. _____

_____

2. _____

_____

3. _____

_____

4. _____

_____

5. _____

_____

6. _____

_____

7. _____

_____

8. _____

_____

9. _____

_____

Define tactical objectives:

_____

_____

List the nine strategic goals and list two tactical objectives that could be used to satisfy each:

1. _____

_____

_____

2. _____

_____

_____

3. _____

4. _____

5. _____

6. _____

7. _____

8. _____

9. _____

Define tactical methods and give examples of tactical methods used to satisfy two of the tactical objectives that you listed above:

_____

For the following descriptions, write a (G) if it is a Strategic Goal, an (O) if it is a Tactical Objective, or an (M) if it is a Tactical Method.

Raising a 24′ extension ladder to a second-story window to rescue a trapped occupant. _____

Vertical ventilation. _____

Advancing a hose line on the unburned portion of a burning house. _____

Performing a right-handed search of a bedroom. _____

Opening a ceiling with a pike pole to seek hidden fire. _____

Using Salvage covers to cover valuables. _____

Assignment of a Rapid Intervention Crew. _____

Assign Incident Safety Officer. _____

Rehabilitation sector and staging sector will maintain accountability boards for crews in their respective sectors. _____

Implement Fireground Accountability System. _____

Arriving units will give passport tag to the accountability location prior to entering hazard zone. _____

Conduct primary search/location. _____

Provide for Medical needs. _____

Evacuation. _____

Assist building occupants from the area as needed and as resources permit. _____

Confinement. _____

Place hose line of sufficient gpm flow on unburned side of the fire. _____

Remove undamaged valuables from the fire area. _____

Describe the need for and the use of the plan of action:

_____

_____

_____

_____

_____

_____

List and describe the fours ways of communicating the plan of action:

_____

_____

_____

_____

_____

_____

_____

_____

Describe the type of occupancies that must be preplanned.

Real World Scenario: You have been given the duty of setting up a preincident planning program at your department. What type of building to you include on the preplan list? What critical items should be included on the list? It will take at least a year to preplan all the needed buildings in your response zone. How do you prioritize the buildings that need to be preplanned most?

_____

_____

_____

_____

Describe the process of recognition-primed decision making:

_____

_____

_____

_____

_____

_____

_____

Chapter

6

Firefighter Safety

LEARNING OBJECTIVES

Upon completion of this chapter, you should be able to:

- Discuss the relationship between firefighter health and safety and strategy and tactics.
- Describe the reports available to study fireground injuries and deaths.
- List the most common causes of firefighter injuries and deaths.
- Describe the impact of regulations and standards on fireground operations.
- List general fireground safety concepts.
- Discuss the strategic considerations for safe fireground operations based on the general safety concepts including, incident safety officers, personnel protective equipment, accountability, rapid intervention crews, and rehabilitation.

Define the following Key Terms:

Occupational Safety and Health Administration

_____

Two-in, two-out rule

_____

Regulations

_____

Standards

_____

Code of Federal Regulations

_____

Fire brigades

_____

Consensus standards

_____

Standard of care

_____

Personnel accountability reports (PAR)

_____

Rapid intervention crews (RIC)

_____

Rehabilitation

_____

Describe the emergency scene risk/benefit philosophy:

_____

_____

_____

_____

Describe the information that is available from the following sources as it relates to firefighter safety and health:

National Fire Protection Association (NFPA)

_____

_____

_____

_____

United States Fire Administration (USFA)

_____

_____

_____

_____

International Association of Firefighters (IAFF)

_____

_____

_____

_____

National Institute for Occupational Safety and Health

_____

_____

_____

_____

Describe in your own words the findings of the NPFA 1998 Injury Report:

_____

_____

_____

_____

_____

_____

_____

_____

_____

_____

Describe in your own words the findings of the USFA's Firefighter Fatalities in the United States 1998:

_____

_____

_____

_____

_____

_____

_____

_____

Describe the impact of regulations and standards on firefighting strategies and tactics:

_____

_____

_____

_____

_____

_____

_____

_____

_____

What three OSHA regulations are applicable to the two-in, two-out ruling:

_____

Real World Scenario: You are the Officer on the engine company that arrives first at a house fire. Heavy smoke and flames are coming out of the kitchen window. Your second due unit is still at least 5 minutes out. Your engine is staffed with a firefighter, an engineer, and yourself. It is 1900 hours. There is a vehicle in the driveway and you notice a child's swing set in the back yard. Your department's SOGs mandates following the two-in, two-out ruling. No one has indicated whether the house is occupied or unoccupied. What action should you take and why? Justify your response.

_____

_____

Explain why a department would want to comply with consensus standards even if they have not been officially adopted by the jurisdiction.

_____

_____

_____

_____

_____

_____

Real World Scenario: You are the Officer on a truck company and the IC orders you and your crew to the roof of a strip mall to perform a trench cut. You understand that the IC has chosen this as a defensive move and that the building will likely be lost entirely if the trench cut is not completed quickly and accurately. Approximately five minutes after you begin cutting, the Incident Safety Officer (ISO) orders you off the roof. The ISO does not go through the IC in this case. What is your course of action and why did you choose it? Justify your answer.

_____

_____

_____

Discuss the need for and the implementation of the following general fireground safety concepts:

Incident Safety Officer

_____

_____

_____

Personal protective equipment

_____

_____

_____

Fireground accountability

_____

_____

_____

Rapid intervention crews

_____

_____

_____

Fire Ground Rehabilitation

_____

_____

_____

Real World Scenario: The ISO orders you to rehab after you have used two bottles of air while fighting a commercial structure fire. You and your crew feel fine and want to continue to work. There are ample resources on the fireground. What is your course of action and why did you choose it?

_____

_____

Chapter

7

Company Operations

LEARNING OBJECTIVES

Upon completion of this chapter, you should be able to:

- List the basic engine company duties.
- Describe responsibilities of the engine company.
- List the basic ladder company functions.
- Describe the support functions of the ladder company.
- List the considerations for apparatus placement.

Define the following Key Terms:

Company

_____

Engine

_____

Booster tank

_____

Capacity hookup

_____

Static supply

_____

Water tender

_____

Ladder

_____

Squad

_____

Rescue

_____

Autoextension

_____

Immediately dangerous to health or life

_____

Scrub area

_____

Collapse zones

_____

List the NFPA requirements for a Class A pumper:

_____

_____

_____

Describe the firefighter's responsibilities to the apparatus:

_____

_____

Describe the following engine company responsibilities:

Water supply (including the various options for moving water to the scene):

_____

_____

_____

_____

_____

Fire attack lines

_____

_____

_____

_____

_____

_____

Locating, confining, and extinguishing

_____

_____

_____

_____

_____

_____

Describe how the stage of the fire affects fire attack:

_____

_____

_____

_____

_____

_____

_____

_____

_____

_____

_____

Describe a fire attack strategy for the following phases of the fire:

Incipient phase

_____

_____

Free-burning phase

_____

_____

Smoldering phase

_____

_____

Describe the role of the ladder company as it relates to overall fireground operations:

_____

_____

_____

_____

_____

_____

Describe the following ventilation strategies:

Vertical ventilation

_____

_____

_____

_____

_____

_____

Horizontal ventilation

_____

_____

_____

_____

_____

_____

Differentiate between the following pairs of terms:

Natural ventilation versus mechanical ventilation

_____

_____

_____

Venting for life versus venting for fire

_____

_____

Discuss the importance of searching, and also include tips for completing the search:

_____

_____

_____

_____

_____

_____

_____

_____

_____

_____

In terms of conducting a forcible entry, list some considerations for:

Doors

_____

_____

_____

_____

_____

_____

_____

_____

_____

Windows

_____

_____

_____

_____

_____

_____

_____

_____

_____

Describe the following ladder company support functions:

Ladder placement

_____

_____

_____

_____

_____

_____

Overhaul

_____

_____

_____

_____

_____

Salvage operations/loss control

_____

_____

_____

_____

_____

_____

List considerations for engine and ladder company apparatus placement:

_____

_____

_____

_____

_____

_____

_____

_____

_____

_____

_____

Chapter

8

Built-In Fire Protection

LEARNING OBJECTIVES

Upon completion of this chapter, you should be to:

- Explain the need for and importance of why built-in fire protection systems are beneficial to building occupants and firefighters.
- Describe what the main water control valve is and what it does.
- Identify the following types of main water control valves and how to determine whether they are in the open or closed position:
  - outside screw and yoke (OS&Y)
  - post indicator valve (PIV)
  - wall post indicator valve (WPIV)
- Given a diagram, locate and identify the fire department connection (FDC).
- Describe the various means by which sprinkler and standpipe systems may be supplied.
- Outline the major components/valves of a sprinkler or standpipe system.
- Identify and explain the operation of pressure-reducing valves found on standpipe systems.
- Explain the differences between the following types of sprinkler systems and how they activate/operate.
  - wet pipe systems
  - dry pipe systems
  - deluge systems
  - preaction system
- Explain the differences between a residential sprinkler system and those installed in commercial facilities.

- Explain the differences and similarities along with the minimum requirements of the following standpipe systems:
  - Class I system
  - Class II system
  - Class III system
- Describe the following types of special extinguishing agents and the hazards associated with each.
  - dry chemical
  - wet chemical
  - carbon dioxide
  - halogenated agents
- Explain the need for fire department support of built-in fire protection systems.
- Outline the minimum items that should be addressed within a standard operating practice for the support of built-in fire protection systems.

Define the following Key Terms:

Main water control valve

_____

Outside screw and yoke (OS&Y)

_____

Post indicator valve (PIV)

_____

Wall post indicator valve (WPIV)

_____

Main drain

_____

Waterflow alarm

_____

Fire pump

_____

Fire department connection (FDC)

_____

Riser

_____

Accelerators

_____

Exhausters

_____

Halon

_____

Saponification

_____

Discuss the components and the operation of the following four types of sprinkler systems:

Wet pipe systems

_____

_____

_____

_____

Dry pipe systems

_____

_____

_____

_____

Deluge systems

_____

_____

_____

_____

Preaction systems

_____

_____

_____

_____

Label the components of the sprinkler system shown in Figure 8-1.

Figure 8-1 *A wet sprinkler system.*

Describe a residential sprinkler system:

_____

_____

_____

_____

_____

_____

_____

_____

Describe the considerations for fire department support of a sprinkler system:

_____

_____

_____

_____

_____

_____

_____

_____

_____

_____

Differentiate between the following types of standpipe systems:

Class I

_____

_____

_____

_____

_____

Class II

_____

_____

_____

_____

_____

Class III

_____

_____

_____

_____

_____

Describe the various pressure-reducing valves:

_____

_____

_____

_____

_____

_____

Describe the necessary fire department support for standpipe systems:

_____

_____

_____

_____

_____

_____

_____

_____

_____

_____

_____

_____

_____

_____

_____

_____

Real World Scenario: You are the company officer on the engine that arrives first at a reported structure fire in a five-story bank building. You have a preincident plan that you referred to en route and noticed that the building has a wet sprinkler and standpipe system with separate FDCs for each system. Upon arrival you see nothing showing; however, building occupants are coming out of the stairwell and report that there is moderate smoke on the third floor. Which system do you charge first and why?

_____

_____

_____

Describe the components and the operation of the following special extinguishing systems:

Carbon dioxide extinguishing systems

_____

_____

_____

_____

Halogenated agent extinguishing systems

_____

_____

_____

_____

Dry chemical extinguishing systems

_____

_____

_____

_____

Wet chemical extinguishing systems

_____

_____

_____

_____

Chapter

9

After the Incident

LEARNING OBJECTIVES

Upon completion of this chapter, you should be able to:

- Define an incident termination plan.
- Discuss the three stages of resource demobilization.
- Discuss the purpose and need for a formal and informal postincident analysis.
- Describe the critical incident stress management system.
- List four types of critical incident stress debriefing.

Define the following Key Terms:

Demobilization

_____

Incident termination

_____

Critical incident stress management (CISM)

_____

First-in, last-out

_____

Standardized apparatus

_____

Postincident analysis

_____

Critical incidents

_____

Peer diffusing

_____

Discuss the considerations related to each of the following three phases of incident termination:

Demobilization

_____

_____

_____

_____

_____

Returning to quarters

_____

_____

_____

_____

_____

Postincident analysis

_____

_____

_____

_____

_____

Describe in detail the purpose of the postincident analysis:

_____

_____

_____

_____

_____

_____

_____

Differentiate between the informal PIA and the formal PIA:

_____

_____

_____

_____

_____

_____

_____

_____

_____

How can information derived from the PIA be used to update procedures:

_____

_____

_____

_____

Describe the critical incident stress management process:

_____

_____

_____

_____

_____

_____

_____

_____

_____

_____

Describe the types of CISM:

_____

_____

_____

_____

_____

_____

_____

_____

_____

_____

_____

Real World Scenario: You have been fighting a large brush fire for several hours. Homes have been endangered but all have been spared due to the effective firefighting. There are several agencies on the scene. You find out that a firefighter from a neighboring fire department has died of a heart attack while fighting the fire. Is CISM needed for this incident? Explain your answer.

_____

_____

_____

Chapter

10

One- and Two-Family Dwellings

LEARNING OBJECTIVES

Upon completion of this chapter, you should be able to:

- Identify the common types of construction for one- and two-family dwellings.
- Describe the inherent life safety problems in these dwellings.
- Describe the hazards associated with firefighting in these structures.
- List the strategic goals and tactical objectives applicable to fires in these structures.

Define the following Key Terms:

Fire stops

_____

Fire load

_____

Describe the characteristics of one- and two-family dwellings:

_____

_____

_____

_____

_____

Describe the characteristics of post and frame construction:

_____

_____

_____

_____

_____

_____

_____

In the space below, draw an example of post and frame construction.

Describe the characteristics of balloon frame construction:

_____

_____

_____

_____

_____

_____

_____

_____

In the space below, draw an example of balloon frame construction.

Describe the characteristics of platform frame construction:

_____

_____

_____

_____

_____

_____

_____

In the space below, draw an example of platform frame construction.

Describe the characteristics of ordinary construction:

_____

_____

_____

_____

_____

_____

_____

_____

In the space below, draw an example of ordinary construction.

Describe the hazards commonly encountered in one- and two-family dwelling fires:

_____

_____

_____

_____

_____

_____

_____

_____

_____

_____

_____

_____

_____

_____

_____

_____

_____

_____

_____

_____

_____

_____

Describe the considerations for the following strategic goals and tactical objectives for one- and two-family dwelling fires:

Firefighter safety

_____

_____

_____

Search and rescue

_____

_____

_____

Evacuation

_____

_____

_____

Exposure protection

_____

_____

_____

Confinement

_____

_____

_____

Extinguishment

_____

_____

_____

Ventilation

_____

_____

_____

Overhaul

_____

_____

_____

Salvage

_____

_____

_____

Describe the strategic goals and tactical objectives that would be considerations in the following situation:

It is reported that there is a fire in the basement of the one-family dwelling shown in Figure 10-1.

Figure 10-1   *A single-family dwelling.*

The time is 17:00 and the month is December. The basement is a finished basement and is used by the family as a recreation room.

Firefighter safety

_____

_____

_____

Search and rescue

_____

_____

_____

Evacuation

_____

_____

_____

Exposure protection

_____

_____

_____

Confinement

_____

_____

_____

Extinguishment

_____

_____

_____

Ventilation

_____

_____

_____

Overhaul

_____

_____

_____

Salvage

_____

_____

_____

Describe the strategic goals and tactical objectives that would be considerations in the following situation:

It is reported that there is a fire in the bedroom of the one-story, single-family dwelling shown in Figure 10-2.

Figure 10-2   *A single-family dwelling.*

The time is 03:00 and the month is July. The fire is showing from the bedroom windows and heavy smoke has spread throughout the house. There are two cars in the driveway.

Firefighter safety

_____

_____

_____

Search and rescue

_____

_____

_____

Evacuation

_____

_____

_____

Exposure protection

_____

_____

_____

Confinement

_____

_____

_____

Extinguishment

_____

_____

_____

Ventilation

_____

_____

_____

Overhaul

_____

_____

_____

Salvage

_____

_____

_____

Describe the strategic goals and tactical objectives that would be considerations in the following situation:

It is reported that there is a fire on the first floor of the two-story, single-family dwelling shown in Figure 10-3.

Figure 10-3   *A single-family dwelling.*

The time is 08:00 on a Saturday in May. The fire started in the first-floor kitchen area, extended to the living room, and is about to extend up the interior stairs to the second floor.

Firefighter safety

_____

_____

_____

Search and rescue

_____

_____

_____

Evacuation

_____

_____

_____

Exposure protection

_____

_____

_____

Confinement

_____

_____

_____

Extinguishment

_____

_____

_____

Ventilation

_____

_____

_____

Overhaul

_____

_____

_____

Salvage

_____

_____

_____

Describe the strategic goals and tactical objectives that would be considerations in the following situation:

It is reported that there is a fire in the attic of the two-story, single-family dwelling shown in Figure 10-4.

Figure 10-4   *A single-family dwelling.*

The time is 01:00 on a weekday in October. The fire occurred after there was a lightning storm in the area. Heavy smoke is showing upon arrival.

Firefighter safety

_____

_____

_____

Search and rescue

_____

_____

_____

Evacuation

_____

_____

_____

Exposure protection

_____

_____

_____

Confinement

_____

_____

_____

Extinguishment

_____

_____

_____

Ventilation

_____

_____

_____

Overhaul

_____

_____

_____

Salvage

_____

_____

_____

Describe the strategic goals and tactical objectives that would be considerations in the following situation:

It is reported that there is a fire in the attached garage of the single-family dwelling shown in Figure 10-5.

Figure 10-5   *A single-family dwelling.*

The time is 21:00 and the month is June. There are two vehicles in the garage at the time of the fire. Both are fully involved, as is the garage itself.

Firefighter safety

_____

_____

_____

Search and rescue

_____

_____

_____

Evacuation

_____

_____

_____

Exposure protection

_____

_____

_____

Confinement

_____

_____

_____

Extinguishment

_____

_____

_____

Ventilation

_____

_____

_____

Overhaul

_____

_____

_____

Salvage

_____

_____

_____

Chapter 11

Multiple-Family Dwellings

LEARNING OBJECTIVES

Upon completion of this chapter, you should be able to:

- Identify the common types of construction for multiple-family dwellings.
- Describe the hazards associated with firefighting in these structures.
- List the strategic goals and tactical objectives applicable to fires in these structures.
- List specific tactical objectives applicable to fires in these structures.

Define the following Key Terms:

Shelter in place

_____

Bulkhead/scuttle

_____

Throat

_____

Trench cut

_____

Describe the characteristics of multi-family dwellings:

_____

_____

_____

_____

_____

Describe the considerations for the following strategic goals and tactical objectives for multi-family dwelling fires:

Firefighter safety

_____

_____

_____

Search and rescue

_____

_____

_____

Evacuation

_____

_____

_____

Exposure protection

_____

_____

_____

Confinement

_____

_____

_____

Extinguishment

_____

_____

_____

Ventilation

_____
_____
_____

Overhaul

_____
_____
_____

Salvage

_____
_____
_____

Describe the construction features and hazards commonly encountered in older apartment houses:

_____
_____
_____
_____
_____
_____
_____
_____
_____
_____
_____
_____
_____
_____
_____
_____
_____
_____
_____

Describe the construction features and hazards commonly encountered in newer apartment houses:

_____

_____

_____

_____

_____

_____

_____

_____

_____

_____

_____

_____

_____

_____

_____

_____

_____

_____

_____

Describe the construction features and hazards commonly encountered in fire-resistive multiple dwellings:

_____

_____

_____

_____

_____

_____

_____

_____

_____

_____

_____

_____

_____

_____

_____

_____

_____

_____

Describe the construction features and hazards commonly encountered in row frame multiple dwellings:

_____

_____

_____

_____

_____

_____

_____

_____

_____

_____

_____

_____

_____

_____

_____

_____

_____

_____

_____

_____

Describe the construction features and hazards commonly encountered in brownstone multiple dwellings:

_____

_____

_____

_____

_____

_____

_____

_____

_____

_____

_____

_____

_____

_____

_____

_____

_____

_____

_____

_____

Describe the construction features and hazards commonly encountered in garden apartments:

_____

_____

_____

_____

_____

_____

_____

_____

_____

_____

_____

_____

_____

_____

_____

_____

_____

_____

_____

_____

_____

Describe the strategic goals and tactical objectives that would be considerations in the following situation:

It is reported that there is a fire on the second floor of the multi-family dwelling shown in Figure 11-1.

Figure 11-1 *A multi-family dwelling.*

The time is 14:00 and the month is January. The fire has fully involved the apartment and is venting out the windows.

Firefighter safety

_____

_____

_____

Search and rescue

_____

_____

_____

Evacuation

_____

_____

_____

Exposure protection

_____

_____

_____

Confinement

_____

_____

_____

Extinguishment

_____

_____

_____

Ventilation

_____

_____

_____

Overhaul

_____

_____

_____

Salvage

_____

_____

_____

Describe the strategic goals and tactical objectives that would be considerations in the following situation:

It is reported that there is a fire on the top floor of the multi-family dwelling shown in Figure 11-2.

Figure 11-2 _A multi-family dwelling._

The time is 18:00 and the month is August. The fire has fully involved the apartment and is venting out the windows and into the hallway. The resident left the apartment door open when he fled.

Firefighter safety

_____

_____

_____

Search and rescue

_____

_____

_____

Evacuation

_____

_____

_____

Exposure protection

_____

_____

_____

Confinement

_____

_____

_____

Extinguishment

_____

_____

_____

Ventilation

_____

_____

_____

Overhaul

_____

_____

_____

Salvage

_____

_____

_____

Describe the strategic goals and tactical objectives that would be considerations in the following situation:

It is reported that there is a fire on the first floor of the multi-family dwelling shown in Figure 11-3.

Figure 11-3   *A multi-family dwelling.*

The time is 16:00 and the month is November. There is heavy smoke throughout the hallway and apartment. No fire is visible.

Firefighter safety

_____

_____

_____

Search and rescue

_____

_____

_____

Evacuation

_____

_____

_____

Exposure protection

_____

_____

_____

Confinement

_____

_____

_____

Extinguishment

_____

_____

_____

Ventilation

_____

_____

_____

Overhaul

_____

_____

_____

Salvage

_____

_____

_____

Describe the strategic goals and tactical objectives that would be considerations in the following situation:

It is reported that there is a fire in the attic of the multi-family, row frame dwelling shown in Figure 11-4.

Figure 11-4   *Row houses.*

The time is 05:00 and the month is December. The fire may have started due to a faulty chimney. There is heavy smoke coming from the attic that extends over multiple units in the building.

Firefighter safety

_____

_____

_____

Search and rescue

_____

_____

_____

Evacuation

_____

_____

_____

Exposure protection

_____

_____

_____

Confinement

_____

_____

_____

Extinguishment

_____

_____

_____

Ventilation

_____

_____

_____

Overhaul

_____

_____

_____

Salvage

_____

_____

_____

Describe the strategic goals and tactical objectives that would be considerations in the following situation:

It is reported that there is a fire in the basement of the multi-family dwelling shown in Figure 11-5.

Figure 11-5 *A multi-family dwelling.*

The time is 02:00 and the month is April. There is heavy smoke and fire visible from the basement on arrival.

Firefighter safety

_____

_____

_____

Search and rescue

_____

_____

_____

Evacuation

_____

_____

_____

Exposure protection

_____

_____

_____

Confinement

_____

_____

_____

Extinguishment

_____

_____

_____

Ventilation

_____

_____

_____

Overhaul

_____

_____

_____

Salvage

_____

_____

_____

Describe the strategic goals and tactical objectives that would be considerations in the following situation:

It is reported that there is a fire on the first floor of the multi-family dwelling shown in Figure 11-6.

Figure 11-6 *Garden apartments.*

The time is 12:00 noon, and the month is February. There is heavy smoke and fire visible from the first floor on arrival.

Firefighter safety

_____

_____

_____

Search and rescue

_____

_____

_____

Evacuation

_____

_____

_____

Exposure protection

_____

_____

_____

Confinement

_____

_____

_____

Extinguishment

_____

_____

_____

Ventilation

_____

_____

_____

Overhaul

_____

_____

_____

Salvage

_____

_____

_____

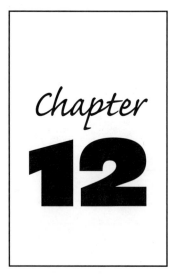

Chapter **12**

Commercial Buildings

LEARNING OBJECTIVES

Upon completion of this chapter, you should be able to:

- Identify the common types of construction for commercial buildings.
- Describe the hazards associated with fire fighting in these structures.
- List the strategic goals and tactical objectives applicable to fires in these structures.
- List specific tactical objectives applicable to fires in these structures.

Define the following Key Terms:

Cantilever

_____

Spalled

_____

Pyrolyzing

_____

Taxpayer

_____

Describe the characteristics of commercial buildings:

_____

_____

_____

_____

_____

Describe the considerations for the following strategic goals and tactical objectives for commercial buildings:

Firefighter safety

_____

_____

_____

Search and rescue

_____

_____

_____

Evacuation

_____

_____

_____

Exposure protection

_____

_____

_____

Confinement

_____

_____

_____

Extinguishment

_____

_____

_____

Ventilation

_____

_____

_____

Overhaul

_____

_____

_____

Salvage

_____

_____

_____

Describe the construction features and hazards commonly encountered in strip centers:

_____

_____

_____

_____

_____

_____

_____

_____

_____

_____

_____

_____

_____

_____

_____

_____

_____

_____

_____

Describe the construction features and hazards commonly encountered in large commercial structures:

_____

_____

_____

_____

_____

_____

_____

_____

_____

_____

_____

_____

_____

_____

_____

_____

_____

_____

_____

_____

_____

Describe the construction features and hazards commonly encountered in two- or three-story commercial structures:

_____

_____

_____

_____

_____

_____

_____

_____

_____

_____

_____

_____

_____

_____

_____

_____

_____

_____

_____

_____

Describe the construction features and hazards commonly encountered in stand alone commercial occupancies:

_____

_____

_____

_____

_____

_____

_____

_____

_____

_____

_____

_____

_____

_____

_____

_____

_____

_____

_____

Describe the strategic goals and tactical objectives that would be considerations in the following situation:

It is reported that there is a fire in the end store of the strip center shown in Figure 12-1.

Figure 12-1 *A strip center.*

The time is 09:00 and the month is September. The fire is in the rear storage area of the store.

Firefighter safety

_____

_____

_____

Search and rescue

_____

_____

_____

Evacuation

_____

_____

_____

Exposure protection

_____

_____

_____

Confinement

_____

_____

_____

Extinguishment

_____

_____

_____

Ventilation

_____

_____

_____

Overhaul

_____

_____

_____

Salvage

_____

_____

_____

Describe the strategic goals and tactical objectives that would be considerations in the following situation:

It is reported that there is a fire in the large commercial structure shown in Figure 12-2.

Figure 12-2   *A commercial building.*

The time is 04:00 and the month is May. The fire is in the center of the building, which is used for manufacturing plastic packing materials. The building is fully fire sprinklered.

Firefighter safety

_____

_____

_____

Search and rescue

_____

_____

_____

Evacuation

_____

_____

_____

Exposure protection

_____

_____

_____

Confinement

_____

_____

_____

Extinguishment

_____

_____

_____

Ventilation

_____

_____

_____

Overhaul

_____

_____

_____

Salvage

_____

_____

_____

Describe the strategic goals and tactical objectives that would be considerations in the following situation:

It is reported that there is a fire on the first floor of the commercial structure shown in Figure 12-3.

Figure 12-3 *A commercial building.*

The time is 14:00 and the month is March. The fire is in an office storage closet. There is heavy smoke on the first floor and light smoke on the second and third floors. The building is fully fire sprinklered.

Firefighter safety

_____

_____

_____

Search and rescue

_____

_____

_____

Evacuation

_____

_____

_____

Exposure protection

_____

_____

_____

Confinement

_____

_____

_____

Extinguishment

_____

_____

_____

Ventilation

_____

_____

_____

Overhaul

_____

_____

_____

Salvage

_____

_____

_____

Describe the strategic goals and tactical objectives that would be considerations in the following situation:

It is reported that there is a fire in the stand alone commercial structure shown in Figure 12-4.

Figure 12-4 *A stand alone commercial building.*

The time is 02:00 and the month is January. The building is fully involved on arrival.

Firefighter safety

_____

_____

_____

Search and rescue

_____

_____

_____

Evacuation

_____

_____

_____

Exposure protection

_____

_____

_____

Confinement

_____

_____

_____

Extinguishment

_____

_____

_____

Ventilation

_____

_____

_____

Overhaul

_____

_____

_____

Salvage

_____

_____

_____

Chapter

13

Places of Assembly

LEARNING OBJECTIVES

Upon completion of this chapter, you should be able to:

- Define what constitutes a place of public assembly.
- Identify the types of construction for places of assembly.
- Understand the inherent life safety problems in these buildings.
- Know the different hazards associated with fire fighting in these structures.
- List the strategic goals and tactical objectives in handling a fire in these structures.

Describe the characteristics of assembly occupancies:

_____

_____

_____

_____

_____

_____

_____

_____

_____

_____

_____

_____

Describe the construction features and hazards commonly encountered in churches:

_____

_____

_____

_____

_____

_____

_____

_____

_____

_____

_____

_____

_____

_____

_____

_____

_____

_____

_____

_____

_____

_____

Describe the construction features and hazards commonly encountered in exhibit halls:

_____

_____

_____

_____

_____

_____

_____

_____

_____

_____

_____

_____

_____

_____

_____

_____

_____

_____

_____

_____

_____

_____

Describe the construction features and hazards commonly encountered in sports arenas:

_____

_____

_____

_____

_____

_____

_____

_____

_____

_____

_____

_____

_____

_____

_____

_____

_____

_____

_____

_____

Describe the construction features and hazards commonly encountered in nightclubs and showplaces:

_____

_____

_____

_____

_____

_____

_____

_____

_____

_____

_____

_____

_____

_____

_____

_____

_____

_____

_____

_____

_____

_____

Describe the considerations for the following strategic goals and tactical objectives for assembly occupancies:

Firefighter safety

_____

_____

_____

Search and rescue

_____

_____

_____

Evacuation

_____

_____

_____

Exposure protection

_____

_____

_____

Confinement

_____

_____

_____

Extinguishment

_____

_____

_____

Ventilation

_____

_____

_____

Overhaul

_____

_____

_____

Salvage

_____

_____

_____

Describe the strategic goals and tactical objectives that would be considerations in the following situation:

It is reported that there is a fire in the church shown in Figure 13-1.

Figure 13-1   *A typical church.*

The time is 03:00 and the month is February. There is heavy smoke showing on arrival; fire is not visible.

Firefighter safety

_____

_____

_____

Search and rescue

_____

_____

_____

Evacuation

_____

_____

_____

Exposure protection

_____

_____

_____

Confinement

_____

_____

_____

Extinguishment

_____

_____

_____

Ventilation

_____

_____

_____

Overhaul

_____

_____

_____

Salvage

_____

_____

_____

Describe the strategic goals and tactical objectives that would be considerations in the following situation:

It is reported that there is a fire in the exhibit hall shown in Figure 13-2.

Figure 13-2   *An assembly occupancy.*

The time is 09:00 and the month is October. There is light smoke throughout that has an electrical smell. There is no visible fire. The fire alarm has been activated. There are about 300 people in the exhibit hall; most are making their way to the exits.

Firefighter safety

_____

_____

_____

Search and rescue

_____

_____

_____

Evacuation

_____

_____

_____

Exposure protection

_____

_____

_____

Confinement

_____

_____

_____

Extinguishment

_____

_____

_____

Ventilation

_____

_____

_____

Overhaul

_____

_____

_____

Salvage

_____

_____

_____

Describe the strategic goals and tactical objectives that would be considerations in the following situation:

It is reported that there is a fire in the small sports arena shown in Figure 13-3.

Figure 13-3   *A sports arena.*

The time is 10:00 and the month is September. It appears that the fire involves trash under the stands. There is no fire alarm system. The bleachers are constructed of wood. Most of the spectators are unaware of the fire.

Firefighter safety

_____

_____

_____

Search and rescue

_____

_____

_____

Evacuation

_____

_____

_____

Exposure protection

_____

_____

_____

Confinement

_____

_____

_____

Extinguishment

_____

_____

_____

Ventilation

_____

_____

_____

Overhaul

_____

_____

_____

Salvage

_____

_____

_____

Describe the strategic goals and tactical objectives that would be considerations in the following situation:

It is reported that there is an odor of smoke in the nightclub shown in Figure 13-4.

Figure 13-4   *A nightclub.*

The time is 23:00 and the month is June. No one has been evacuated and there is nothing showing on arrival. As you enter the club, you smell a strong odor of burning wood.

Firefighter safety

_____

_____

_____

Search and rescue

_____

_____

_____

Evacuation

_____

_____

_____

Exposure protection

_____

_____

_____

Confinement

_____

_____

_____

Extinguishment

_____

_____

_____

Ventilation

_____

_____

_____

Overhaul

_____

_____

_____

Salvage

_____

_____

_____

Chapter

14

High Rise Office Buildings

LEARNING OBJECTIVES

Upon completion of this chapter, you should be able to:

- Define high-rise office buildings.
- Describe various construction methods.
- Describe hazards associated with these structures.
- Describe the building systems present in high-rise buildings that can be of use to firefighters.
- Describe the strategic goals and tactical objectives related to high-rise building fires.

Define the following Key Terms:

Plenums

_____

Central (or center) core construction

_____

Dampers

_____

Access stairs

_____

Stack effect

_____

Stratification

_____

Describe the construction common in high-rise office buildings:

_____
_____
_____
_____
_____
_____
_____
_____
_____
_____
_____
_____
_____
_____
_____
_____
_____
_____
_____
_____
_____

Describe the operation and the relationship to the fire fighting of the follow systems found in high-rise office buildings:

Standpipes

_____
_____
_____
_____
_____

Fire pumps and hydraulics

_____

_____

_____

_____

_____

_____

Heating, ventilation, and air conditioning systems (HVAC)

_____

_____

_____

_____

_____

Describe the hazards encountered in high-rise office buildings:

_____

_____

_____

_____

_____

_____

_____

_____

_____

_____

_____

_____

_____

_____

_____

_____

_____

_____

_____

_____

Describe the considerations for the following strategic goals and tactical objectives for high-rise office buildings:

Firefighter safety

_____

_____

_____

Search and rescue

_____

_____

_____

Evacuation

_____

_____

_____

Exposure protection

_____

_____

_____

Confinement

_____

_____

_____

Extinguishment

_____

_____

_____

Ventilation

_____

_____

_____

Overhaul

_____

_____

_____

Salvage

_____

_____

_____

Describe the strategic goals and tactical objectives that would be considerations in the following situation:

It is reported that there is a fire on the 7th floor of the office building shown in Figure 14-1.

Figure 14-1   _A high-rise office building._

The time is 15:00 on a weekday in February. There is heavy smoke showing on the fire floor; the fire is confined to a single office area. The building is of central core construction.

Firefighter safety

_____

_____

_____

Search and rescue

_____

_____

_____

Evacuation

_____

_____

_____

Exposure protection

_____

_____

_____

Confinement

_____

_____

_____

Extinguishment

_____

_____

_____

Ventilation

_____

_____

_____

Overhaul

_____

_____

_____

Salvage

_____

_____

_____

Chapter

15

Vehicle Fires

LEARNING OBJECTIVES

Upon completion of this chapter, the reader should be able to:

- Describe the common types of vehicle fires.
- Describe the hazards associated with fires involving passenger cars, vans, and light duty trucks.
- Describe the hazards associated with fires involving semi-tractor trailers and trucks.
- Describe the hazards associated with fires involving recreation vehicles.
- Describe the hazards associated with fires involving heavy construction equipment.
- Describe the hazards and challenges associated with fires involving buses.
- Apply strategic goals and tactical objectives to these incidents.
- Apply various tactical methods utilized for these incidents.

Describe the characteristics and hazards common to cars, vans, and light duty pickup trucks:

_____

_____

_____

_____

_____

_____

_____

_____

Describe the considerations for the following strategic goals and tactical objectives for vehicle fires in cars, vans, and light duty pickup trucks :

Firefighter safety

_____

_____

_____

Search and rescue

_____

_____

_____

Evacuation

_____

_____

_____

Exposure Protection

_____

_____

_____

Confinement

_____

_____

_____

Extinguishment

_____

_____

_____

Ventilation

_____

_____

_____

Overhaul

_____

_____

_____

Salvage

_____

_____

_____

Describe the characteristics and hazards common to semi-tractor trailer and truck fires:

_____

_____

_____

_____

_____

_____

_____

_____

Describe the considerations for the following strategic goals and tactical objectives for semi-tractor trailer and truck fires :

Firefighter safety

_____

_____

_____

Search and rescue

_____

_____

_____

Evacuation

_____

_____

_____

Exposure protection

_____

_____

_____

Confinement

_____

_____

_____

Extinguishment

_____

_____

_____

Ventilation

_____

_____

_____

Overhaul

_____

_____

_____

Salvage

_____

_____

_____

Describe the characteristics and hazards common to fires in recreational vehicles:

_____

_____

_____

_____

_____

_____

_____

Describe the considerations for the following strategic goals and tactical objectives for fires in recreational vehicles :

Firefighter safety

_____

_____

_____

Search and rescue

_____

_____

_____

Evacuation

_____

_____

_____

Exposure protection

_____

_____

_____

Confinement

_____

_____

_____

Extinguishment

_____

_____

_____

Ventilation

_____

_____

_____

Overhaul

_____

_____

_____

Salvage

_____

_____

_____

Describe the characteristics and hazards common to fires in heavy construction equipment:

_____

_____

_____

_____

_____

_____

_____

_____

Describe the considerations for the following strategic goals and tactical objectives for fires in heavy construction equipment :

Firefighter safety

_____

_____

_____

Search and rescue

_____

_____

_____

Evacuation

_____

_____

_____

Exposure protection

_____

_____

_____

Confinement

_____

_____

_____

Extinguishment

_____

_____

_____

Ventilation

_____

_____

_____

Overhaul

_____

_____

_____

Salvage

_____

_____

_____

Describe the characteristics and hazards common to fires in buses:

_____

_____

_____

_____

_____

_____

_____

_____

Describe the considerations for the following strategic goals and tactical objectives for fires in buses:

Firefighter safety

_____

_____

_____

Search and rescue

_____

_____

_____

Evacuation

_____

_____

_____

Exposure protection

_____

_____

_____

Confinement

_____

_____

_____

Extinguishment

_____

_____

_____

Ventilation

_____

_____

_____

Overhaul

_____

_____

_____

Salvage

_____

_____

_____

Describe the strategic goals and tactical objectives that would be considerations in the following situation:

It is reported that there is a fire at the construction site shown in Figure 15-1.

Figure 15-1 *Heavy construction equipment.*

The time is 09:00 on a weekday in May. There is heavy smoke and fire showing from the back hoe motor.

Firefighter safety

_____

_____

_____

Search and rescue

_____

_____

_____

Evacuation

_____

_____

_____

Exposure protection

_____

_____

_____

Confinement

_____

_____

_____

Extinguishment

_____

_____

_____

Ventilation

_____

_____

_____

Overhaul

_____

_____

_____

Salvage

_____

_____

_____

Describe the strategic goals and tactical objectives that would be considerations in the following situation:

It is reported that there is a fire in the camper shown in Figure 15-2.

Figure 15-2 *A recreational vehicle.*

The time is 12:00 noon on a weekday in July. There is heavy smoke and fire showing from the camper.

Firefighter safety

_____

_____

_____

Search and rescue

_____

_____

_____

Evacuation

_____

_____

_____

Exposure protection

_____

_____

_____

Confinement

_____

_____

_____

Extinguishment

_____

_____

_____

Ventilation

_____

_____

_____

Overhaul

_____

_____

_____

Salvage

_____

_____

_____

Chapter

16

Wildfires

LEARNING OBJECTIVES

Upon completion of this chapter, you should be able to:

- Describe what is considered a wildland/urban interface area.
- Identify the three types of wildland fires.
- Identify the hazards associated with the various wildland fires.
- List the strategic goals and tactical objectives applicable to wildland fires.

Define the following Key Term:

Wildland/urban interface

_____

Describe the characteristics of the wildland/urban interface:

_____

_____

_____

_____

_____

_____

_____

_____

_____

_____

Describe the characteristics and hazards encountered with forest fires:

_____

_____

_____

_____

_____

_____

_____

_____

_____

_____

_____

_____

_____

_____

_____

_____

_____

_____

_____

_____

_____

_____

Describe the characteristics and hazards encountered with brush fires:

_____

_____

_____

_____

_____

_____

_____

_____

_____

_____

_____

_____
_____
_____
_____
_____
_____
_____
_____
_____
_____
_____

Describe the characteristics and hazards encountered with grassland fires:

_____
_____
_____
_____
_____
_____
_____
_____
_____
_____
_____
_____
_____
_____
_____
_____
_____
_____
_____
_____
_____

Describe the strategic goals and tactical objectives that would be considerations in wildland fires:

Firefighter safety

_____

_____

_____

Search and rescue

_____

_____

_____

Evacuation

_____

_____

_____

Exposure protection

_____

_____

_____

Confinement

_____

_____

_____

Extinguishment

_____

_____

_____

Ventilation

_____

_____

_____

Overhaul

_____

_____

_____

Salvage

_____

_____

_____

Describe the strategic goals and tactical objectives that would be considerations in the following situation:

It is reported that there is a fire in the wildland/urban interface shown in Figure 16-1.

Figure 16-1 _A wildland/urban interface._

The time is 09:00 on a Saturday in June. There is heavy smoke and fire showing in the wooded area.

Firefighter safety

_____

_____

_____

Search and rescue

_____

_____

_____

Evacuation

_____

_____

_____

Exposure protection

_____

_____

_____

Confinement

_____

_____

_____

Extinguishment

_____

_____

_____

Ventilation

_____

_____

_____

Overhaul

_____

_____

_____

Salvage

_____

_____

_____

Chapter

17

Special Fires (Basic Information)

LEARNING OBJECTIVES

Upon completion of this chapter, you should be able to:

- Describe the various types of special situations that firefighters may encounter.
- Describe the hazards associated with fires involving hazardous materials.
- Describe the hazards associated with fires involving ships.
- Describe the hazards associated with fires involving railcars.
- Describe the hazards associated with fires involving aircraft.
- Describe the hazards associated with fires involving bulk storage facilities.
- Apply strategic goals and tactical objectives to these incidents.

Define the following Key Terms:

Detonate

_____

Deflagrate

_____

Fire control plan

_____

BLEVEs

_____

Describe the hazards associated with response to hazardous materials incidents:

_____

_____

_____

_____

_____

_____

_____

_____

_____

_____

_____

_____

_____

_____

_____

_____

_____

_____

_____

_____

_____

Describe the strategic goals and tactical objectives that would be considerations in hazardous materials fires:

Firefighter safety

_____

_____

_____

Search and rescue

_____

_____

_____

Evacuation

_____

_____

_____

Exposure protection

_____

_____

_____

Confinement

_____

_____

_____

Extinguishment

_____

_____

_____

Ventilation

_____

_____

_____

Overhaul

_____

_____

_____

Salvage

_____

_____

_____

Describe the hazards associated with response to shipboard fires:

_____

_____

_____

_____

_____

_____
_____
_____
_____
_____
_____
_____
_____
_____
_____
_____
_____
_____
_____
_____
_____
_____
_____

Describe the strategic goals and tactical objectives that would be considerations in shipboard fires:

Firefighter safety

_____
_____
_____

Search and rescue

_____
_____
_____

Evacuation

_____
_____
_____

Exposure protection

_____
_____
_____

Confinement

_____

_____

_____

Extinguishment

_____

_____

_____

Ventilation

_____

_____

_____

Overhaul

_____

_____

_____

Salvage

_____

_____

_____

Describe the hazards associated with response to railcar fires:

_____

_____

_____

_____

_____

_____

_____

_____

_____

_____

_____

_____

_____

_____

_____

_____

_____

_____

_____

_____

_____

Describe the strategic goals and tactical objectives that would be considerations in railcar fires:

Firefighter safety

_____

_____

_____

Search and rescue

_____

_____

_____

Evacuation

_____

_____

_____

Exposure protection

_____

_____

_____

Confinement

_____

_____

_____

Extinguishment

_____

_____

_____

Ventilation

_____

_____

_____

Overhaul

_____

_____

_____

Salvage

_____

_____

_____

Describe the hazards associated with response to aircraft fires:

_____

_____

_____

_____

_____

_____

_____

_____

_____

_____

_____

_____

_____

_____

_____

_____

_____

_____

Describe the strategic goals and tactical objectives that would be considerations in aircraft fires:

Firefighter safety

_____

_____

_____

Search and rescue

_____

_____

_____

Evacuation

_____

_____

_____

Exposure protection

_____

_____

_____

Confinement

_____

_____

_____

Extinguishment

_____

_____

_____

Ventilation

_____

_____

_____

Overhaul

_____

_____

_____

Salvage

_____

_____

_____

Describe the hazards associated with response to bulk storage facilities:

_____

_____

_____

_____

_____

_____

_____

_____

_____

_____

_____

_____

_____

_____

_____

_____

_____

_____

_____

_____

Describe the strategic goals and tactical objectives that would be considerations in bulk storage facilities:

Firefighter safety

_____

_____

_____

Search and rescue

_____

_____

_____

Evacuation

_____

_____

_____

Exposure protection

_____

_____

_____

Confinement

_____

_____

_____

Extinguishment

_____

_____

_____

Ventilation

_____

_____

_____

Overhaul

_____

_____

_____

Salvage

_____

_____

_____

Describe the strategic goals and tactical objectives that would be considerations in the following situation:

It is reported that there is a fire in the ship shown in Figure 17-1.

Figure 17-1 *A cruise ship.*

The time is 19:00 on a Saturday evening in June. The ship is at dock. There is heavy smoke and fire showing from the deck area.

Firefighter safety

_____

_____

_____

Search and rescue

_____

_____

_____

Evacuation

_____

_____

_____

Exposure protection

_____

_____

_____

Confinement

_____

_____

_____

Extinguishment

_____

_____

_____

Ventilation

_____

_____

_____

Overhaul

_____

_____

_____

Salvage

_____

_____

_____

Describe the strategic goals and tactical objectives that would be considerations in the following situation:

It is reported that there is a fire at the bulk storage area shown in Figure 17-2.

Figure 17-2 *Bulk storage area.*

The time is 13:00 on a Weekday evening in November. There is heavy smoke and fire showing from the Loading Dock.

Firefighter safety

_____

_____

_____

Search and rescue

_____

_____

_____

Evacuation

_____

_____

_____

Exposure protection

_____

_____

_____

Confinement

_____

_____

_____

Extinguishment

_____

_____

_____

Ventilation

_____

_____

_____

Overhaul

_____

_____

_____

Salvage

_____

_____

_____

Chapter

18

Putting It All Together

LEARNING OBJECTIVES

Upon completion of this chapter, you should be able to:

- Apply fireground management concepts, strategic goals, and tactical objectives to a simulated fire in a single-family dwelling.
- Apply fireground management concepts, strategic goals, and tactical objectives to a simulated fire in a multiple-family dwelling.
- Apply fireground management concepts, strategic goals, and tactical objectives to a simulated fire in a commercial building.

Describe the strategic goals and tactical objectives that would be considerations in the following situation:

It is reported that there is a fire in the single-family dwelling shown in Figure 18-1.

Figure 18-1 *A single-family dwelling*

The time is 19:00 on a weekday in January. There is heavy smoke, but no fire showing on arrival.

Firefighter safety

_____

_____

_____

Search and rescue

_____

_____

_____

Evacuation

_____

_____

_____

Exposure protection

_____

_____

_____

Confinement

_____

_____

_____

Extinguishment

_____

_____

_____

Ventilation

_____

_____

_____

Overhaul

_____

_____

_____

Salvage

_____

_____

_____

Describe the strategic goals and tactical objectives that would be considerations in the following situation:

It is reported that there is a fire in the multi-family (taxpayer) building shown in figure 18-2.

Figure 18-2 _A multi-family dwelling._

The time is 04:00 on a weekday in July. There is heavy smoke throughout the building, and fire is visible from one first-floor window.

Firefighter safety

_____

_____

_____

Search and rescue

_____

_____

_____

Evacuation

_____

_____

_____

Exposure protection

_____

_____

_____

Confinement

_____

_____

_____

Extinguishment

_____

_____

_____

Ventilation

_____

_____

_____

Overhaul

_____

_____

_____

Salvage

_____

_____

_____

Describe the strategic goals and tactical objectives that would be considerations in the following situation:

It is reported that there is a fire in the warehouse shown in Figure 18-3.

Figure 18-3 *A commercial building.*

The time is 14:00 on a weekend in May. There is heavy smoke throughout the building, but no fire is visible.

Firefighter safety

_____

_____

_____

Search and rescue

_____

_____

_____

Evacuation

_____

_____

_____

Exposure protection

_____

_____

_____

Confinement

_____

_____

_____

Extinguishment

_____

_____

_____

Ventilation

_____

_____

_____

Overhaul

_____

_____

_____

Salvage

_____

_____

_____

Real World Scenario: It is your first day as a company officer on an engine company. You have been assigned an engineer and a firefighter. You have been friends with the engineer for many years and you know he does excellent work. The firefighter has been with the department only 6 months and you do not know her well. It is shift change time and the off-going officer just handed you the keys as the alarm sounds for a single-family structure fire. An elderly man is trapped in the building. You are first due and 2 minutes away from the fire. Starting right now (before you even leave the station), what is going through your mind?

_____

_____

_____